The Chinese Sailing Rig

Design and Build Your Own Junk Rig

BY

Derek Van Loan

WITH ILLUSTRATIONS BY AUTHOR

Copyright © 2007 text and illustrations by Derek Van Loan

All rights reserved. No part of this book may be used or reproduced in any manner whatsoever without written permission, except in the case of brief quotations embodied in critical articles and reviews. For information, contact the publisher.

Illustrations by Derek Van Loan
Cover design by Rob Johnson, www.robjohnsondesign.com
Book design by Linda Morehouse, www.webuildbooks.com
Cover photo and photo on p. vii © copyright 1983 Mark Balsiger,
    mbalsiql@elp.rr.com

Printed in the United States
Second Edition

ISBN 0-939837-70-6, 978-0-939837-70-6

Published by
Paradise Cay Publications, Inc.
P. O. Box 29
Arcata, CA 95518-0029
800-736-4509
707-822-9163 Fax
paracay@humboldt1.com
www.paracay.com

My wife of thirty years, Mae Margaret, who typed the original manuscript and the manuscript for the second printing and worked with me on all my projects, passed away nearly four years ago. She would be pleased to know that her energy continues to help people.

Contents

Introduction

    Take a Hammer   3
    Scale   4
    Auxiliary Rigs   4
    Modeling   5
    Western Materials   5

Section One

    A Practical Sail   11
    How Many Battens?   12
    Laying it Out   12
    Sail Area   14
    How Many Masts?   15
    Built-in Headwind   16
    Choosing a Course of Action   16
        To Determine the CLR of a Vessel   18
        To Determine the CE of a Vessel   19
        To Determine the Combined CE   20
        To Find the CE of More Than Two Sails   21
        To Determine the Lead   21
    Where to Place the Masts?   22
    Preliminary Drawing   23
    Sleep on it   25

Section Two

    Spars   29
    Dimensions   29
    Formula   30
    Taper   32
    Partners   32
    Bury   33

Section Two (Cont'd)

Practical Partners	35
Wedges	36
Keeping Water Out	37
Mast Step	38
Obtaining a Timber Mast	40
Shaping the Mast	40
Mast Finish	44
Masthead Fitting	45
Masthead Accessories	47

Section Three

Theory and Practice	51
What Size Yard, Boom?	53
Sail Materials	54
Attaching Battens	57
Two Methods of Rigging	58
Halyards	58
Lifts	59
Parrels	63
Sheets	64
Reefing Downhauls	69

Under Way

Jibing	75
Seeing the Wind	75
Don't Forget	76
Easy Reefing	77
A Good Idea?	77
Spaghetti	78
Advantages	79

The Decked Sailing Canoe
    History   83
    My Canoe   84
    Where to Get Sailcloth   85
    Method   85
    Fittings   86
    Masts   86
    Battens   87
    Quick Rig   87
Reading List   91
    The Internet   95
    Periodicals   95
Glossary   97

Lorca

FERROCONCRETE, BUILT BY HOWARD VAN LOAN
48' LOA, BEAM 13.5', DRAFT 6', SAIL AREA 1,200 SQ. FT.
FOREMAST 45', 10" DIA. @ PARTNERS
MAINMAST 51', 12" DIA. @ PARTNERS

PHOTO © 1983 MARK BALSIGER

The Chinese Sailing Rig

INTRODUCTION

INTRODUCTION

This guide has grown out of years of designing, building and sailing Chinese lug rigs. With this information my fellow builders and I have been able to actualize our dreams. Much impetus for this text was provided by my lack of spare time for answering letters of inquiry about Chinese rigs. Also, each time I loaned out my notes, they were returned in a slightly more tattered condition. This work is derived from these notes, which were the result of my experience with this rig. I want to say a big thanks to my brother Howard for copyediting this latest edition. All I know about this subject is contained here. You may come to different conclusions. Please see the reading list for more information.

TAKE A HAMMER IN ONE HAND AND...

When I began designing and building Chinese lug rigs in the 1960s, not much information was available. There were a few magazine articles, all with inadequate detail, and some historically oriented books, but no real hammer-and-nails approach to the subject. There is more information today, thanks to The Junk Rig Association and to magazine coverage. The field, however, is still wide open.

SCALE

The specifications for gear and fittings presented here are rather on the heavy side. For lighter displacement vessels, aluminum and stainless steel fittings can be fabricated that are lighter than the galvanized steel ones I generally use. You might want to look into tapered aluminum flag poles for masts. They are tough, light, and available. Wooden wedges at the partners are used with metal masts as well as the solid wood masts presented here. Hollow wood spars for the Chinese rig constructed using Barry Noble's system, also look very good. Metal fittings are not absolutely necessary: Chinese boatbuilders used lashings, bamboo, and wood fittings extensively.

AUXILIARY RIGS

Perhaps, as fuel becomes more expensive, auxiliary sailing rigs will begin to sprout on powerboats. Some of the older powerboats are especially adaptable to sail. The Chinese lug sail would be a good choice for those powerboat owners who are thinking about obtaining a little "silent power" from the wind. This rig can be controlled from the pilot house and is essentially a low aspect ratio rig. Power boaters considering a sailing rig might bear in mind that a larger rudder—or better still, an auxiliary rudder hung on the transom—is almost a necessity. A two-masted rig can be made to help with steering, and, since it keeps the center of pressure lower, might be better than a single-masted auxiliary rig.

MODELING

I can't recommend too strongly that the amateur designer make a model of his Chinese lug rig before building full size. Dowels can be used for the mast, boom, and yard. Thin sticks can be used for battens. Glue the yard, boom, and battens onto light cloth. Use screw eyes for blocks and string for rigging. You can mount the rig by plugging your dowel-mast into a hole drilled in a two-by-four.

EASTERN AND WESTERN MATERIALS

When I first started to construct my own rigs, I thought it was necessary to use bamboo for battens. I soon discovered that, although bamboo is excellent, it is hard to find where I live. I found it more reasonable to utilize indigenous materials such as Douglas fir in my rig construction. The Chinese sail was originally put together using grass mats, and was designed to distribute the load evenly over the entire rig. Western sailing rigs tend to concentrate load stresses, and Western technology has produced stronger materials that can withstand them. Bear in mind these basic philosophies as you design your rig.

SECTION ONE

The Chinese Sailing Rig

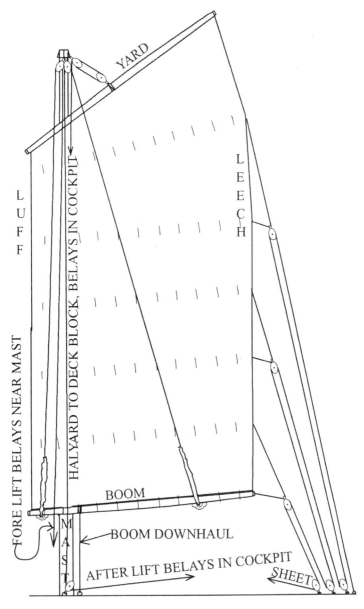

DRAWING I: PORT SIDE
(SEE DRAWINGS 26 & 27 FOR REEFING)

A Practical Sail

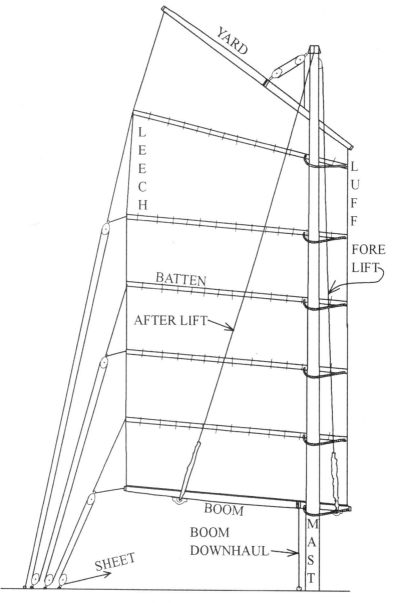

DRAWING 2: STARBOARD SIDE
(SEE DRAWINGS 26 & 27 FOR REEFING)

THE CHINESE SAILING RIG

A PRACTICAL SAIL

The Chinese lug sail is a fully battened standing lug sail (as opposed to a dipping lug, which is pulled around to the other side of the mast when the vessel tacks). The Chinese lug sail is a quadrilateral sail, as is the gaff sail, but it is not a square sail (squar'sl) as is sometimes thought. Square sails normally bear the wind on only one side, whereas the Chinese lug sail bears the wind on both sides.

There are various profiles—high and low peaks, etc.—but the profile that seems to work best for me, as far as reefing, furling, and set go, is a variation of the Chinese ocean-going junk sail. I once thought that a high peaked sail, having an effectively higher aspect ratio, would go to windward significantly better. I constructed my first sails in this manner. I found that the high peak is hard to sheet without producing undue twist, and the peak, because it is in the wind shadow of the yard, actually contributes little to windward efficiency.

Dave Mallory (a San Francisco Bay Area friend) has worked out my favorite profile for the Chinese lug sail. He used it for years on his converted 36-foot lifeboat *Kokoro*, and I adapted it for our 46-foot schooner *Yankee Belle*.

Start with a peak angle of 125 degrees (with the luff) and equal-length battens, yard, and boom. Make the tack angle 85 to 90 degrees (to the luff) and make all the lower battens parallel with the boom (foot of the sail). To determine the distance between each lower batten, divide the luff height by the total number of battens. Place the top batten one-quarter of the lower batten interval below the yard at the luff.

HOW MANY BATTENS?

The number of battens is dependent on the sail area. Up to 200 square feet, four or five battens will suffice. Use five or six battens on a sail where the area is from 200 to 700 square feet. This batten/sail area ratio works well, giving correct camber to the airfoil and support to the sail fabric. Also, the sheet purchase power is adequate using this guide. The traditional grass mat Chinese sail had extra battens, with no sheets attached, for additional strength.

LAYING IT OUT

On a line from peak to clew, measure peak to intersection of the uppermost of the lower battens. Divide this distance in half and run the uppermost batten from its luff point through this halfway mark to the leech. This actually determines where the leech is at the upper batten. Remember all the battens are of equal length. The leech runs in straight lines from batten to batten.

Ideally, the foot of the sail should be about .66 to 1.2 times the length of the luff. This will produce a reasonable aspect ratio, even

A Practical Sail

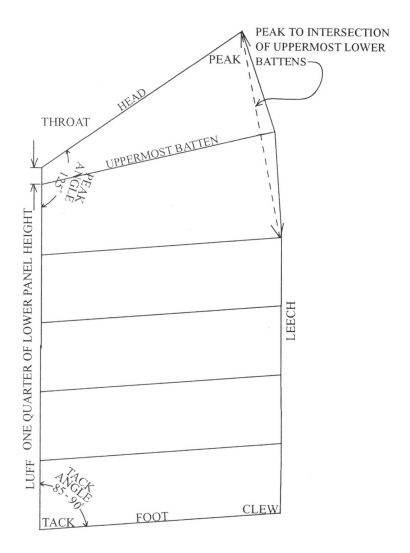

DRAWING 3
SAIL PROFILE AND BATTENS

with two reefs down. The full sail will appear rather lofty, but the sail plan should incorporate enough area to drive the vessel in light airs without necessitating the setting of another sail. The big advantage of the Chinese lug rig is the ease with which you can reduce sail, without having to thrash around on a wet, slippery deck.

SAIL AREA

You should know how to find the area, in square feet (or meters, etc.) of a sail. There are various methods used by yacht designers. For our purposes a method requiring no special tools is all that's necessary. One method is to draw the sail on graph paper and count the squares inside the outline. Another method is to divide the sail into squares, oblongs, and triangles for which you find the areas, all of which you add together.

To determine the total sail area for the vessel we must consider what a similar vessel would have carried before the days of auxiliary engines. *American Small Sailing Craft* by H. I. Chapelle, Eric Hiscock's *Cruising Under Sail*, and *Voyaging Under Sail* are a few well documented guides. Maritime museums can be a great help. Another guide is the comparative sail area carried on similar modern craft including light-air sails, but excluding spinnakers, as it is impossible to design a practical Chinese lug rig that could incorporate this much area. Our vessel, *Yankee Belle*, has approximately 150 square feet more sail area than the original pre-engine 1890s design. A satisfactory amount of sail area should probably contain more than just the equivalent of main, mizzen, and genoa alone. If conditions warrant, you can sail with a reef or two.

HOW MANY MASTS?

The mast or masts should stand perpendicular to the water line to prevent the sail from winging inboard in light airs. Raking the masts forward a little is quite acceptable, because having the sails swing out in a calm is desirable. Forward rake is aesthetically displeasing to some people, though. The sail normally overlaps the mast with 25 percent or less of its area. Less is better. The mast is roughly parallel to and aft of the sail's luff. The lower halyard block attaches at or near the center of the yard. The designer must be careful to allow for the height (distance) along the mast taken up by halyard blocks and tackle. Allow extra mast height for shackles, boom and yard thickness, and for masthead fittings. Allow a little extra mast height so that the sail could be moved forward to help compensate for excessive weather helm. Allow, too, a few inches for stretch of the sail fabric. When the sail is set, the boom should clear the cockpit by a safe margin, too. It is important that there be enough room for the sheets to be pulled in tight when the sail is fully furled and down on the gallows. See Drawing 23.

For a number of reasons, on any vessel of more than about 25 feet overall length, I prefer a two-masted Chinese lug rig. Since the rig is heavy, the two-masted configuration helps keep the weight and the center of pressure (CE) low. It is a lot easier to maneuver and balance a two-masted rig because the CEs exert their force nearer the ends of the vessel, and for this same reason any design idiosyncrasies that might appear in sailing can be more easily compensated for. When running wing-and-wing, the two-masted rig tends to keep the vessel from rounding up as it would with a single sail set out to one side. Also, the two-masted rig allows you to set a light sail such as a golly wobbler between the masts. Disadvantages of the two-masted rig are the extra weight, duplication of gear, and the slight inconvenience of more line handling. I still prefer the two-masted rig, and I have sailed with both.

BUILT-IN HEADWIND

A two-masted Chinese lug schooner is aerodynamically more like a sloop without an overlapping jib than a schooner. This is because, like a sloop, there are only two sails that are normally set. A conventional schooner has a mainsail, foresail, and a jib, and perhaps a fore staysail. Therefore, the Chinese lug schooner has the advantage of less aerodynamic sail interference, which makes it analogous to a biplane rather than a triplane or quadraplane. That is, it has less built-in headwind.

I use the term "schooner" to describe a rig with two masts and two Chinese sails of the same size or with a larger after sail. If the after sail were smaller I would call it a ketch or a yawl, depending on the placement of the mizzenmast relative to the rudder post. A yawl or ketch's after sail is called a "mizzen."

CHOOSING A COURSE OF ACTION

Once the approximate sail area is determined, much trial and error is required to properly adapt the new rig to the vessel. From Column A of Table 1, select one of the four options that best matches the information you have. Then move to Column B and follow the directions.

At this stage of your design, some method of self steering must be considered, for without it the vessel will require constant, tedious attention on a long passage. If your boat will steer herself you will find this a useful feature, even for day sailing. Presently the two distinct methods for making a sailing vessel self steer are: 1) to balance the sail area using sheet-to-tiller arrangements (See Donald Ridler's excellent drawings in his *Eric the Red*) and 2) to use a self-steering device.

The self-steering device can be an electric auto pilot or a wind vane self-steering gear. With the latter, the vane must be kept clear of the sheets and the sail. Therefore, right from the beginning of the design process you must decide whether to have a self-steering device, and you must decide what type to use.

TABLE I
TO DETERMINE CE/CLR
CHOOSE A COURSE OF ACTION FROM COLUMN B
GIVEN THE INFORMATION YOU HAVE IN COLUMN A

COLUMN A IF YOU HAVE...	COLUMN B
CE From profile design drawing	Use this to design new Chinese rig (assuming that existing rig is correctly balanced).
CLR From profile design drawing	Use this to estimate CE using percent of waterline length and suggestions for lead given in text.
A profile design drawing of your hull.	Use this to calculate CLR and estimate, using percent of waterline length and lead measurements suggested in text, to determine position of the CE.
A profile design drawing of existing rig	Use profile design to calculate CE of existing rig and then use this information to design the Chinese rig.

TABLE 1: TO DETERMINE THE CLR OF A VESSEL

A scale representation of the underwater profile, excluding the rudder, should be drawn on stiff paper. The outline should be cut out (folded lengthwise a little if necessary to keep the ends from drooping) and balanced on a pin until it is level. Press the pin into the paper to mark the CLR. See Drawing 4 below.

If the rudder of the vessel is large or if the vessel has a fin keel, daggerboard, or centerboard, include about one-third its area, from its leading edge, in your underwater profile.

Draw a line perpendicular to the waterline through the CLR on the profile design drawing of the hull.

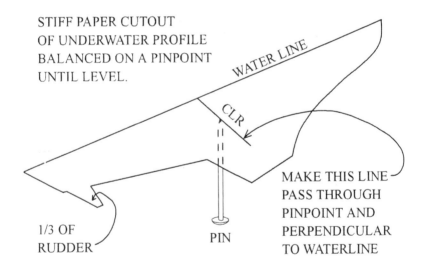

DRAWING 4
CENTER OF LATERAL RESISTANCE

A Practical Sail

TABLE 1: TO DETERMINE CE OF A VESSEL

Determine the CE of each sail, then find the combined CEs of all the sails.

To determine the CE of a sail, draw its profile, to the same scale you used for the hull, on stiff paper, and as with the calculation of the hull's CLR, cut it out and balance it on a pin until it is level. Press the pin through the paper to mark the sail, and label this point "CE."

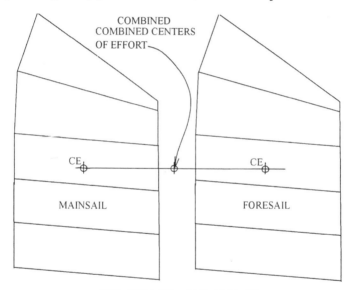

$$\frac{\text{SAIL AREA OF}}{\text{MAINSAIL}} = \frac{\text{SAIL AREA OF}}{\text{FORESAIL}}$$

THE COMBINED CENTERS OF EFFORT IS HALFWAY BETWEEN THE CENTER OF EFFORT OF THE FORE SAIL AND THE CENTER OF EFFORT OF THE MAIN SAIL

DRAWING 5
COMBINED CENTERS
EQUAL AREA SAILS

TABLE 1: **TO DETERMINE THE COMBINED CE**

Determining the combined CE of all the sails of a vessel is done mathematically. Using the scale profile design drawing, draw a line through the CE of any two sails. Using dividers or a ruler, find the distance between the two CEs. Divide the product of multiplying the area of the foresail and the scaled distance between the CE of each sail, by the sum of the areas of the two sails. See Drawing 6 below.

This will give you the distance, from the after sail, along the line connecting the two centers, to the combined CE. Mark this combined CE on the line.

Of course with two sails of the same shape and area, the combined CE falls exactly halfway between the CE of each sail. See Drawing 5 on the preceding page.

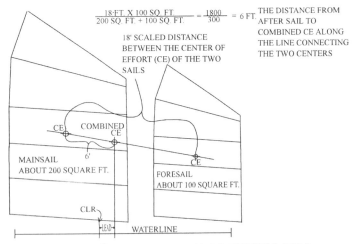

DRAWING 6: COMBINED CENTERS, UNEQUAL SAILS

TABLE 1: TO FIND THE CE OF MORE THAN TWO SAILS

Find the combined CE of any two sails and draw a line through this point and the CE of the adjacent sail. Use the above method (for finding the combined CE of any two sails) to find the combined CEs, treating the combined CE of the two sails as one sail. Repeat this until all the sails in the rig are taken into account.

TABLE 1: TO DETERMINE THE LEAD

A sailing vessel in the water is like a weather vane on a barn. The sailboat pivots around its CLR just as the weather vane has its pivot. The weather vane has most of its sail area (CE) aft of its pivot point so that it points into the wind. A properly designed sailing vessel has its sail area close to its CLR so that it can be sailed on any course in respect to the wind by altering the CE. On a sailing vessel we alter the CE by sheeting the sail in or out and by setting or reducing sail.

The concept of CE and CLR is convenient and arbitrary; we do not really know at any given time where the real CLR and CEs are as the vessel makes its way through the varying forces of water and wind. But the CE–CLR concept produces well-balanced sailing vessels.

All this gets us to the determination of positioning the CE in relation to the CLR. Lead (see glossary) is expressed as a percentage of the vessel's waterline length. Keel yachts have been found to work well with a lead of about .015 to .15 lead. That is with the CE .015 to .15 of the waterline length *ahead* of the CLR. This means that a vessel with a 30-foot waterline might work well with a designed CE three feet ahead of the designed CLR along the waterline.

Centerboard vessels can vary the CLR by changing the position of their board. They have been found to work well with a lead of .03 to .10.

In general, a higher rig produces more weather helm because as a vessel heels under pressure of the wind, the sail that is pulling the vessel forward tends also to pull the vessel around. So we should give a vessel with a tall rig a bit more lead forward of the CLR to compensate.

When you decide on the amount of lead to give your rig, draw a line on the scale drawing perpendicular to the waterline at the point on the waterline that you have determined to be the proper lead. Label this line CE.

WHERE TO PLACE THE MAST(S)

With the two-masted Chinese lug rig, I treat the foresail as I would the jib of a conventional cutter rig. The mast should go forward, the step being even as far forward as the forwardmost end of the waterline. If there is not enough room on the foredeck to handle anchors and tackle because part of the sail forward of the mast sweeps this area, the anchor windlass can be placed alongside the foremast, because the foresail generally weathercocks when one is anchoring or weighing anchor.

The aftermast (main in a schooner, mizzen in a ketch) is stepped somewhat aft of the middle of the vessel. If there is adequate buoyancy forward in the hull, this mast arrangement works well and you can make both sails the same area.

A PRACTICAL SAIL

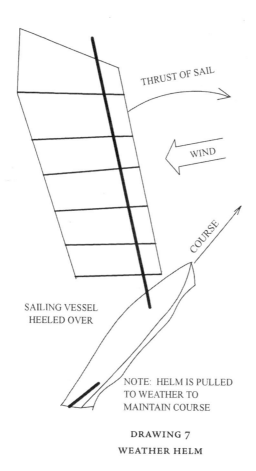

DRAWING 7
WEATHER HELM

PRELIMINARY DRAWING

Let's begin with two equal-sized sails. To the same scale as your hull profile, draw two sails (or one if you decided on a single mast), each half of your total sail area, on a sheet of tracing paper. Use the instructions for laying out the Chinese lug sail profile given in Drawing 3.

First be sure to allow space along the deck to accommodate the sails and sheets aft of their leeches, and for enough distance between the sails to prevent contact. Allow enough space between sails so that fore and aft adjustments can be made later during sailing trials. For a big rig allow at least three feet between the leech of the foresail and the luff of the main.

You will probably want to use standard lumber dimensions for these spars. Remember that this is a trial-and-error process and that you will have to make several paper sails before you will be satisfied.

Mark the mast on the sails and place the sails on the hull profile with the CLR on it. Move the sails on the drawing to determine mast placement. Do not worry if things do not look as if they will work out; you might need to design a boomkin to sheet the main (after) sail to, or change the proportions of the sail. Bear in mind that a full bulkhead or other substantial reinforcement will be needed forward or aft of the masts, and that sliding hatch covers on the centerline will need room to slide. Speaking of hatches, do not step masts near any large deck opening or the partners' supports may interfere with them.

If your design using equal-sized sails does not seem to be working out, a change in the relative sizes of the sails may be necessary, and you will have to compute the combined center of effort, which you then place over the correct lead just as you did with the equal-sized sails. (Remember that the combined CE for two equal-sized sails falls halfway between their individual centers.) A vessel with a long, skinny bow—that is, one having a small amount of buoyancy forward—should have a smaller sail forward, or the forward sail should be stepped well aft of the bow, as this sail may depress the bow excessively. Rigs with two sails generally work well with sails that range from being equal in area to those where one is twice the area of the other.

SLEEP ON IT

If things seem to be working out, be skeptical. Make a few tentative design changes. You might find an even better arrangement. Do not hurry this stage of the design: "Sleeping on it" certainly applies here when you do come to an acceptable compromise. The extreme options generally include the choice of making the rig too lofty, or of having sails that are too stumpy—that is, with sails whose aspect ratio is too low—or in having to carry excessively reefed sail area.

You must decide how tall a rig your vessel can carry; when you reef a sail, its center of pressure and its weight of the sail come down. But masts can't be reefed and they contribute to a vessel's heeling. This is one of the most difficult decisions to make, and here again you must go by what similar and especially traditional vessels with solid spars carry. The Chinese mast has a weight aloft advantage over the gaff mast, in that it is tapered from deck to masthead, whereas gaff rig masts are parallel-sided to the hounds and therefore are heavier aloft for the same mast height carrying the same sail area. Whatever you decide, there is a practical maximum mast length. Remember, too, you can always cut a rig down, but it is a lot harder to add to it. Properly set up, a good strong boomkin is a really good way to add length to a vessel that is short and beamy. To avoid snagging the sheets in a jibe, locate the lower sheet blocks inboard from the boomkin's tip about six inches or so. With a little thinking, snipping out of paper sails, and some sliding them around, you should be able to come up with as good a sailing rig as that of any designer. Try to imagine all the operations that you would go through in sailing the vessel, and how your design will affect them. Take an imaginary sail from mooring to dock, out to sea and back again.

Again, if you design a single-sail rig, follow the same procedure, except for finding the combined CE.

SECTION TWO

SPARS

The Chinese lug rig, without stays and shrouds for mast support, has been in daily (although declining) commercial use for two to three thousand years. Yet to many in the West, a sailing rig without any standing rigging is a radical concept. But modern wire standing rigging is a relative latecomer to the rigs of the small sailing craft. Some of the advantages of the unstayed rig are:

1) Reduction of sail chafe, especially when sailing off the wind

2) No expensive standing rigging, and the fittings associated with it

3) Reduced windage with the added advantage of less wind noise in a breeze

4) Less maintenance

5) Less compression strain on the mast.

DIMENSIONS

The main factors that influence the dimensions of the mast are sail area, relative stiffness (stability) of the hull, the length of the mast, and mast material and quality.

No formula for determining the diameter at the partners should be followed blindly. All the factors above and perhaps others should be taken into consideration in unstayed mast design. After all, the ultimate authority is the sea.

There are three factors in the formula I use for calculating mast diameter (in inches) at the partners.

A = number of square feet of sail

H = height of mast above the partners in inches

S = safety multiplier:

    1.5 - 2 for small, lightweight vessel

    2 for small ballasted vessel

    3 for large vessels

    3.5 for beamy vessels

FORMULA

For any lug-rigged vessel over 35 feet LOA, I would use a safety multiplier (S) of 3.5. A really stiff vessel should have a slightly larger mast diameter at the partners. The results produced by this formula will seem huge for anyone familiar with wire-stayed mast sizes, but one soon gets used to having stays and shrouds incorporated into the mast.

SPARS

FORMULA FOR MAST DIAMETER AT PARTNERS

$$\text{Diameter in inches} = \sqrt[3]{\frac{16\,A\,H\,S}{15700}}$$

This means: The diameter of the mast at the partners is equal to the cube root of 16 times A, times H, times S. What you get when you multiply these figures together is divided by 15700. You then calculate the cube root of that number. You now have the mast diameter at the partners.

If you would like to find the diameter at the partners of a fir mast that is 24 feet (288 inches) tall above the partners and is to be stepped in a small, ballasted, single-masted vessel, with a planned 200 square feet of sail and a safety multiplier of 2, set up the problem in the following manner:

1) $\text{Diameter in inches} = \sqrt[3]{\frac{16 \text{ times } 288 \text{ times } 200 \text{ times } 2}{15700}}$

2) $\text{Diameter in inches} = \sqrt[3]{\frac{1843200}{15700}}$

3) $\text{Diameter in inches} = \sqrt[3]{117.4}$

4) Diameter in inches = about 5 inches

TAPER

Make a straight taper from a little above the partners to the masthead, which should be approximately .5 x diameter at the partners. This means that the diameter at the masthead is .5 times (one half) the diameter at the partners. I make the masthead diameter equal to the next larger outside diameter of standard pipe or tube size, because the masthead fitting is fabricated of steel pipe or tubing and it is easier to buy standard-sized pipes. The masthead fitting bears on a shoulder formed by carefully cutting away the wood with a chisel and driving the fitting into place. See Drawing 14, page 54.

PARTNERS

The force of the wind on the sail and the inertial forces of a vessel rolling in the sea must be distributed into and absorbed by the hull and deck of the sailing vessel. With an unstayed rig these forces are distributed solely through the partners and step. It is highly advisable to have a full or a least a partial bulkhead at each mast. If you intend to step the mast through the cabin top, a full or partial bulkhead or heavy knees, preferably steel, must be built into the vessel. Masts should not be stepped athwart large hatches. If no easy way of providing reinforcement is obvious, you might construct reinforcement above deck. The Chinese did this with external deck beams, incorporating deck stowage below gratings that rested on the beams.

Designing partners is relatively simple once the diameter of the mast is determined. (See Drawing 8, page 42.) A set of wedges made of wood that is softer than the mast (such as cedar or white pine) is fitted shoulder-to-shoulder around the mast, and these are contained by a steel tube. The surface area of the partners must be large enough to prevent crushing of wood fibers. The wedges should be longer than the depth of the partner tube, and they should be cut with an overhanging "hook" so that if they loosen they will not fall through the tube and into the vessel. Alternatively, a block of wood could be added to a wedge to provide the "hook."

BURY

The amount of "bury"—that is, the depth between the mast step and the partners—must be determined. Too little bury makes for a disproportionately large amount of leverage force that has to be absorbed by step and partners. If you have any difficulty understanding these forces, take a pencil and using your fingers as partners and step, vary the "bury" and twist the pencil around to simulate the forces applied to a mast. Of course, in general, a larger vessel will have greater depth from deck to step. In a vessel in which this depth is small, the partners might, with suitable bracing, be entirely above deck. The minimum depth from partners to step should range from .1 x mast length to .15 x mast length. Of course, smaller vessels can have proportionately less bury, but large vessels must have more. Masts longer than forty feet should not go below .12 x mast length or 4.8 feet unless they are very light and carry little sail.

As for surface area of the partners, make it as large as possible. A rough guide is shown in the following table.

THE CHINESE SAILING RIG

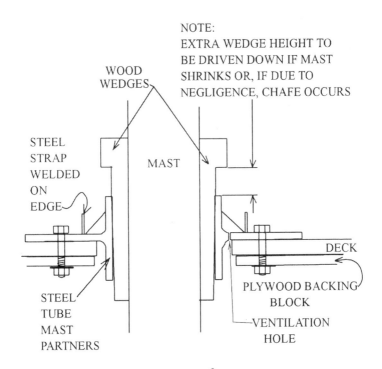

DRAWING 8
CROSS SECTION STEEL MAST PARTNERS

TABLE 2
TO DETERMINE STEEL TUBE THICKNESS OF PARTNERS

Mast Length (in feet)	Minimum Height of Partners Tube (in inches)	Rough Guide to Steel Thickness (in inches)
20	4	3/16 - 1/4
30	6	1/4
40	8	1/4
50	12	3/8

PRACTICAL PARTNERS

The mast partners can be welded of mild steel, galvanized, and bolted into the vessel. The steel pipe or structural tube is gusseted, welded into a piece of plate, and bolted on the deck with plywood backing blocks and steel washers on the underside. This makes a strong sandwich that is in contact with a lot of deck.

Be sure to determine the proper fore and aft angle of the tube before welding. Remember, the mast should be either raked forward or perpendicular to the waterline. The angle can be found by using a stout stick (15' to 20' long) as a temporary mast. Do a lot of sighting of all angles from a dinghy, preferably with the help of two assistants. String stays will help here.

Deck-block attachments can be cut out of ¼" – ⅜" steel plate drilled for shackles, ground smooth, and welded to the partners' plate on either side of the mast. Provide enough space for shackles and blocks, and a couple of extra holes that might come in handy later. Gusset the attachment fittings and build up a good weld fillet.

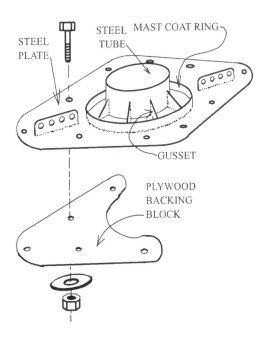

DRAWING 9
STEEL MAST PARTNERS

These attachments should be designed so that the lines leading aft to the cockpit will be high enough to clear deck houses, opening hatches, and other obstacles, or designed to fair lead around them.

WEDGES

The wedges should be cut with a very slight taper so that they can be driven really tight. They should be checked periodically as they can loosen up easily. Several temporary wedges can be driven around the mast to secure it while the permanent ones are carved. It takes some time to fit the wedges because they must be repeatedly trimmed and tried until they all fit, leaving enough wedge to be driven in later if the mast shrinks.

The steel ring around the wedges should be large enough to allow for good sized wedges which can be easily driven with a big hammer. This means that the ring should be approximately three inches in diameter larger than the mast at the partners. After the wedges "settle in" a strip of thin wood or metal can be wrapped around them and a small nail or screw can be driven through into each wedge. (No nails into the mast, please!) The wedges will still have to be checked periodically. See Drawing 8, page 42.

KEEPING WATER OUT

The mast coat is seized onto the mast, or clamped on using stainless (all stainless) hose clamps (pieced together if necessary) with much good bedding compound or good quality caulk. The bottom goes around the steel band that is welded on edge onto the partners' plate. The mast coat should be sewed together and waterproofed. The seizing or clamp around the mast can be hidden under the coat, if the top of the coat is put on first with the outside of the coat to the mast, then turned down to fit at the deck. See Drawing 9 on page 44 and Drawing 10 on page 46.

MAST STEP

The mast step, too, is best fabricated of steel. It should be bolted to as much of the vessel as practical in order to distribute the large forces of the mast. In a wood vessel at least three floor timbers should be spanned. The best attachment for the step may be the cast ballast. If a new vessel with concrete ballast is being rigged, a portion of the step might be cast into the ballast (being careful to leave drainage for the heel of the mast.). I stepped my 50' masts by first suspending them at the partners using temporary wedges to hold the steps. Then I poured the ballast concrete around the steps that extend into the keel. This allowed me to make the final adjustments in mast rake.

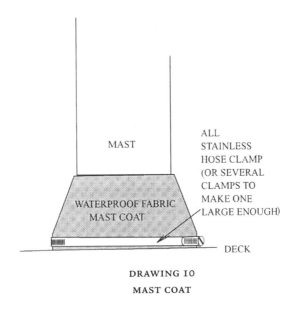

DRAWING 10
MAST COAT

A steel box, open on top and bottom, is welded of heavy plate (¼" or thicker) with a slight taper and steel brackets for bolting it to the vessel. If you are casting it into the ballast, cast re-bar through holes in the step. A hole should be drilled through the sides of the box fore and aft to receive a pin that will secure the mast when it is stepped. Alternatively, a cylindrical steel tube section can be tightly fitted and driven on with a sledge hammer. A fore-and-aft pin should be used here also. See Drawing 11 below.

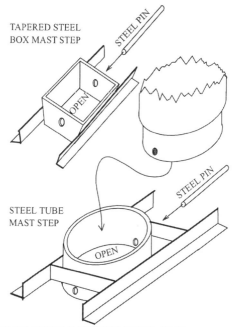

LEAVE BOTTOMS OPEN FOR VENTILATION
TO HELP PREVENT ROT IN MAST

DRAWING 11
MAST STEPS

OBTAINING A TIMBER MAST

I have used Douglas fir almost exclusively, but you may be able to find another suitable wood. Probably the best way to choose your mast stick is to go to your local power or telephone company's source of poles. It is here that the selection of a straight, close grained, small knot, unchecked or small checked, rot-free mast stick can best be made. Try to get an untreated stick. Although "dry treatment" sticks work okay, you should use an air-fed helmet or an appropriate mask when cutting them. Treated poles stink up a closed vessel. Do not use a creosoted (black greasy) stick. If the public utility company source is not available, you might try a wooden-flagpole maker, and, as a last resort, your local forest. Wrap and store your stick under cover so that the sun and wind will not dry it out too quickly and cause checking. A pole yard is a better place than a forest to choose a stick because it is easier to see the details, mentioned above, that make for a good mast.

SHAPING THE MAST

A fairly easy method of shaping a mast out of a grown stick was introduced to me by my friend Dennis Nowlen and has been used with improvements in technique by myself and others to shape several 30- to 50-foot Douglas fir masts.

It goes like this:

1) First draw a scale representation of the stick that you have, using, say, one-half inch to the foot. Take diameter measurements

(take two sets if the stick is oval) every three feet along the stick. You can make your own pair of calipers of three-eighths-inch plywood, a bolt, washers, and a wingnut.

2) Draw a scale representation of the mast that you want inside the first drawing, centering the two so they are exactly superimposed.

Make the mast taper towards the masthead slightly at the partners where the mast wedges will bear. Make a straight line taper from the top of the partners section to the masthead diameter (which is .5 x partner diameter). This gives us a diagrammatic guide for the depths we have to cut. See Drawing 12 below.

DRAW A SCALE REPRESENTATION OF THE MAST AND THE STICK
WITH CENTERS EXACTLY SUPERIMPOSED

DRAWING 12
LAYING OUT MAST SHAPE

3) Now, to determine how much to cut off the stick to produce a straight taper from the top of the partners cylindrical section to the masthead, we must proportionally figure out the diameter of the mast at three-foot intervals or less. For example, say you have a stick that is 10 feet long (partners to head), and you have determined that you need a twelve inches diameter at the partners. You want six

inches at the head. Further calculating the mast taper, you will find that the mast should be nine inches in diameter halfway between the top of the partners and the head. It follows that the mast should be 10.5 inches in diameter ten feet above the partners, and so forth. Check your diagram with careful mathematical figuring of these dimensions along the stick. The mast should be carefully measured at the points that you have determined and marked with a wax crayon. Transfer all this information to the scale diagram so that you can keep it organized. Use this diagram to double-check your math, too. Below the partners, the mast may be tapered down to the step.

4) Set the circular saw to ½ of the difference between the diameters of the stick and the mast, minus about ⅛" to allow for error. Double-check calculated dimensions against the scale diagram and always double-check the saw setting before making cuts. Take your time. Cutting a mast is like cutting a diamond out of the rough.

5) Using an electrical handheld circular saw (preferably worm-drive type and preferably with a carbide blade) make ring (annular) cuts around the stick at five-inch intervals along the stick with the saw set to the depth determined in step 3. Use depth measurements from your diagram to set and reset your saw as you move along the mast making cuts around it. Hold the blade guard back when starting the cut, and wear goggles, ear protection, and mask. It helps to have someone roll the stick for you, too. Make several cuts through each knot and as much as possible, keep the center of the blade in a line with the center of the stick and the saw base tangent to the circumference. See Drawing 13 next page.

6) Sharpen an ax, and standing alongside the stick, carefully chip the scored surface down to the rough mast dimensions. (This

is really good exercise.)

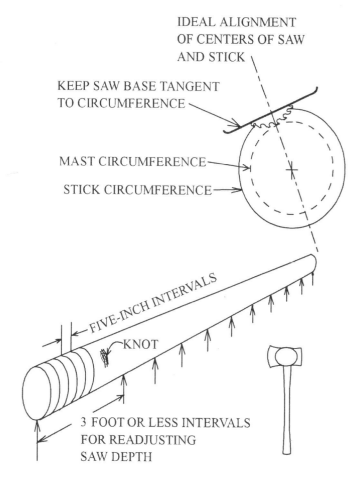

DRAWING 13
SHAPING A GROWN STICK

7) A little work with a draw knife, and now the mast can be planed by hand or with a big power hand planer and belt sanded.

Working alone, using this method and a power planer, I have easily shaped two 50-foot sticks in less than a week. Yards and booms can be done at the same time.

This method of carving down a grown stick into a mast seems considerably easier than what I have observed of the traditional method where one first squares a round stick, then makes it eight-sided, sixteen-sided, etc., and finally planes it round again. A few words of caution: With either method it is impossible to straighten out a stick that has grown crooked, as the tensile fibers that you shave off the bump side will not be there to keep the stick from bending further after shaping. Of course if you are starting with an already squared timber, use the traditional method and begin with eight-siding it.

MAST FINISH

Some sort of protection is needed to keep the stick from drying out too fast and checking. I've used paint, linseed oil, and tallow. Paint is probably best. Tallow is messy but effective. Linseed oil does not work very well. Storing the mast on end will help it dry evenly. Stored on its side, it should be well supported and should be rotated periodically. If you do oil your mast, re-oil it frequently as it will dry faster than you might think. Do not putty longitudinal checks or they will trap moisture and rot will occur beneath the putty. According to authorities, longitudinal checks do not affect the mast's strength appreciably.

Leave the ends of the mast untreated until the mast is dry. This will enable the moisture to leave the center of the mast as well as the surface, helping to reduce checking due to uneven shrinking.

MASTHEAD FITTING

The fitting at the masthead can be welded of mild steel and galvanized. Before galvanizing, have the fitting sandblasted to remove the welding slag. A section of pipe with lugs cut from angle iron (because it is tapered and makes for lighter lugs) and drilled for shackles is simple and works well. Of course, weld the thicker section of the angle iron to the masthead fitting. Make the pipe section long enough to provide a good bearing surface to the mast. Weld on brackets for radar reflector, strobe light, antennas, and the like. Avoid creating pockets that will trap moisture—the galvanizers will refuse to plate a fitting that can't be perfectly dried, and trapped moisture will shorten the service life of the fitting.

Carefully carve the mast to fit the masthead fitting exactly, leaving a shoulder to fit against the bottom edge of the fitting. Using a handsaw with a wood depth guide C-clamped to it, saw around the mast at the shoulder. Set the gauge to make a cut equal to the thickness of the steel of the fitting. Then use a sharp chisel and mallet to carve away the wood from the mast tip to the shoulder. The masthead fitting should be driven on, and then a liberal amount of thin epoxy resin poured into the masthead. Git-Rot works well. See Drawing 15, Boom or Yard End Fitting, page 54.

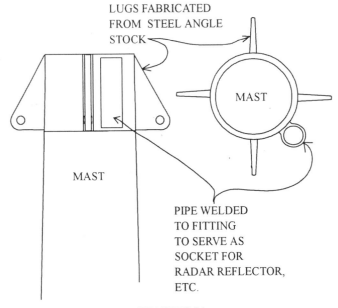

DRAWING 14
MASTHEAD FITTING

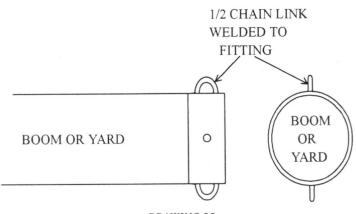

DRAWING 15
BOOM OR YARD END FITTING

Steel is a good material for making sheet attachments, too. They can be as simple as a section of galvanized angle iron with a hole for each sheet block, and mounting holes for bolting to the deck. 316 stainless steel is also quite good for fittings that take a lot of wear and would be liable to rust if made out of mild steel. I use 308 rod for AC arc welding with 316 steel.

MASTHEAD ACCESSORIES

Wires for lightning protection, antennas, and lights can be led loosely down the mast inside the batten parrels. I have done this with hard insulated wires.

Do not lead foam insulated coaxial cable in this manner as it may be crushed, causing it to lose efficiency. Some VHF and ham radio antennas use "co-ax" with a hard plastic core.

For my HF band amateur radio station, I use a length of telephone "drop" wire, with the pair of wires soldered together on each end and terminating at the masthead with the standard phone company strain-relief fitting. This wire is copper-coated spring steel with tough insulation. I use a random length wire antenna tuner to resonate my antenna to the desired frequency.

The antenna, lightning rod, or radar reflector can also serve to keep birds from landing on the masthead. The radar reflector is most effective if mounted in the "raincatcher" position. In the absence of these, a sharp spike at the masthead will help to discourage birds.

Another method for leading wires down the mast is to rout a groove down from the masthead. If you have a hollow mast, the choice is obvious.

SECTION THREE

THEORY AND PRACTICE

The airfoil shape of the Chinese lug sail is created by the bending of the battens. Battens that are not stiff enough, because they are fabricated from a too flexible material, or are cut too thin, or are too widely spaced, will produce a baggy sail shape. See Drawing 16. This will be okay in a light breeze for windward work, but when the wind strengthens, the vessel will not go well to windward. A sail made like this will produce a characteristic S-bend when it lies against the windward side of the mast.

It is better to have battens that are too stiff than too slinky. Of course, as the wind increases, any batten arrangement will begin to bend considerably. Reefing the sail will tend to flatten it out again.

The traditional Chinese sail uses several bamboo poles to make up one batten. Their overlaps are placed in the center of the sail to keep it stiff and flat.

Materials for battens should be locally obtainable. I have used spruce, Douglas fir, bamboo and PVC plastic. (The plastic was unsatisfactory.) Aluminum and fiberglass have been used by others with success, but I have no experience with these.

The battens are a big factor in the weight of the sail. The sheeting arrangement pretty much determines the number of battens, which are evenly spaced. A larger sail requires more sheet purchase, therefore more battens and blocks. A smaller sail requires less purchase. It is possible to put in battens which have no sheet attachment points, but with strong modern fabric this is unnecessary. Finally, round off the edges: A router with a ¼"–⅜" round-off bit works well for this.

DRAWING 16
BATTENS NOT STIFF ENOUGH

TABLE 3: SPRUCE OR FIR BATTEN DIMENSIONS

(BATTENS MAY BE SHAPED LATER)

Batten Length (in feet)	Batten Depth (in inches)	Batten Thickness (in inches)
5	1⅝	½
10	1⅝	1
15	2	1⅝
20	3	1⅝

TABLE 4

SOLID FIR OR SPRUCE YARD DIMENSIONS

Yard Length (in feet)	Diameter at Center (in inches)	Diameter at Ends (in inches)
5	2+	2
10	3	2½
15	4	3½
20	5	4½

WHAT SIZE YARD, BOOM?

Boom diameter can be as great as the diameter of the yard or can be slightly greater than the batten cross-section. The boom should be thick enough not to bow as shown in Drawing 17.

Both yard and boom diameters might be altered a bit at the ends to allow a standard-sized pipe or steel tube to be used in making the end fittings. The end fittings should be drilled for a lagscrew before galvanizing.

THE CHINESE SAILING RIG

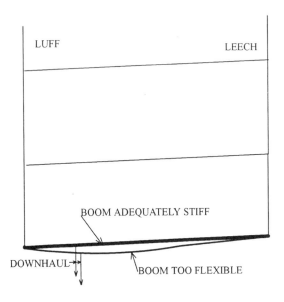

DRAWING 17
BOOM STIFFNESS

SAIL MATERIAL

Sailcloth can be any fabric from grass mats to synthetic fiber cloth. Dacron (Terylene, Tetoron) works best. Cotton is heavy, especially when wet, gets baggy, requires more care, and is almost as expensive as Dacron. Sails have been made from plastic tarp material. I understand the silver tarps are more durable than the blue.

Sailrite (see Reading List) offers a complete line of supplies and tools for amateur sailmakers, everything from fabric to sewing machines. They are reliable and easy to deal with. I used to use pins for sewing. Now I use their Seamstick Basting Tape. It makes sewing cloth together much easier.

The fabric can be lightweight because the Chinese lug sail consists virtually of several smaller sails. Thus, the area of unsupported cloth is small. Here is an approximate guide for sail cloth weight in ounces (American and British) for the total number of square feet of cloth in the sail.

TABLE 5

SAILCLOTH WEIGHT / SAIL AREA

Sail Area in Square Feet	Ounces per Square Yard British	Ounces per Square Yard American
100	5	4
300	6	5
500	8	7

The Dacron sail may be made with doubled-over Dacron tape all the way around. A wide chafing cloth should be sewn to the sail at each batten. Bear in mind that the batten is sandwiched between the sail and the mast. Large multiple corner patches should be sewn at peak and throat. See Drawing 18 following page.

Because the Chinese sail airfoil shape derives solely from the battens, the sailcloth is cut flat. This makes it quite easy for the amateur sailmaker to sew his own sails.

Vertical seams are the best, as the sail will still function if it rips along its seams. Horizontal seams are acceptable, however. Large grommets with liners should be sewn in at peak, throat, tack, and clew. Smaller grommets are used at batten ends, and may be used at

THE CHINESE SAILING RIG

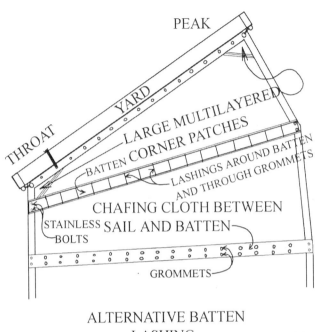

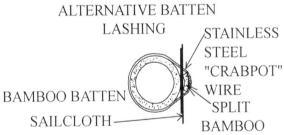

DRAWING 18

BATTEN ATTACHMENT

about one-foot intervals along the head of the sail for yard robands or lacing. Grommets may also be used elsewhere on the sail. Continuous spiral lacing may not be as failsafe as individual lashings (robands) for attaching the head of the sail to the yard. The individual ends of the lashings should be seized with marline or tarred nylon seine twine. Fusing the ends with heat can also help to keep lashings from loosening. See Drawing 18 facing page.

ATTACHING BATTENS

Battens may be tied to the sail with individual lashings. Batten pockets on Chinese lug sails can be problematic. The pockets make batten replacement at sea difficult. Battens may also be attached to the sail by stitching directly through the fabric with marline at about six-inch intervals. A crude but effective method is to wire around a split bamboo pole, which is on the side of the sail opposite the batten. Use stainless steel, "crab pot" wire. Wire the battens on through the sail cloth. It works. Donald Ridler used this method on his boat, *Eric The Red*. He learned of it studying contemporary Chinese junks. See Drawing 18 facing page.

Regardless of the method of batten attachment, the sail must be properly stretched along the battens. The knobs on the bamboo of the original Chinese sails spread the outhaul load evenly along the batten. I have used stainless steel bolts (two on each end) bolted through the batten ends with the sail hauled out along the batten. After the bolts are passed through batten and sail, and tightened down with washers under the heads and nuts, the battens are then cut flush with the luff and leech of the sail. The "sheetlets" or "crows feet" are tied with a tops'l sheetbend or bowline through grommets at each batten at the leech. Tarred nylon seine twine makes excellent small lashings and seizings and lasts longer than marline.

TWO METHODS OF RIGGING

There are two ways of rigging the sail to the mast. One is to rig the sail to its spars aboard the vessel; the other to rig it to its spars ashore.

The first way requires a calm day to bend the sail to the yard that has been previously lashed to the lower halyard block. My wife and I did this with two of our boats, with one of us on each side of the sail, passing the lashings back and forth. As each batten was attached, we raised the sail a bit and lashed on the next batten, until finishing with the boom.

The second method (for strong people if it's a big sail) is to find a large flat space, lay out the sail, lash on all spars, bundle it all up, and take it aboard, where it is lashed to the halyard block.

HALYARDS

Halyards may be 3-strand Dacron line. For easy pulling and long halyard life, the blocks should be a little bit bigger than the rope size. That is, ½" line should run through blocks designed for ⅝" or even ¾" line. Steel-reinforced wood blocks are heavy but work well for halyards. The bare wood cheeks can be soaked for a month in raw linseed oil and dried for two weeks. This reduces maintenance to a minimum. Block bearings must be greased now and then with a good marine grease.

Up to two hundred square feet of sail may be hoisted easily with a three-part purchase: that is, with a double block aloft and a single block lashed to the yard. For two hundred to five hundred square feet, a four-part purchase is required, and for five hundred to eight hundred square feet a five-part purchase. The halyards must be belayed so that they can be hauled and veered from a comfortable position. Considerable effort is expended in raising a large sail. I have found that sitting on the deck with my legs braced against a coaming is most efficient, especially when the vessel is rolling. For any sail above two hundred fifty square feet in area, I use ½" diameter halyard, not so much for strength but for comfortable pulling.

The lower halyard block is lashed to the center of the yard with several round turns to keep it from slipping along the yard. Its fore and aft position on the yard can be adjusted later if necessary. Next, the head of the sail is stretched between the outhaul fittings of the yard and lashed tight, after which the robands or lashing (light Dacron lines) are passed twice through the grommets around the head of the sail and around the yard, and tied. Seize the ends of the robands, then fuse them with heat so that they will not untie. The battens and boom are similarly lashed on; that is, outhauls are pulled tight, then lashed along the sail. See Drawing 21.

LIFTS

The lifts should be rigged before rigging the sail to the mast so that they will be ready to support the furled sail. There are two main lifts for each sail, one forward of the mast and one aft. The lifts should be positioned far enough from the ends of the boom so that the yard cannot get on the wrong side of the lifts. Screw a U-shaped fairlead to the boom over each lift to keep it from sliding along the spar toward the mast. You can make a wood fairlead for this purpose. They are less hazardous than sharp metal fittings.

The lifts are really important to the operation of the Chinese lug sail, so make them thick enough to be strong and to prevent chafing. You can wrap synthetic carpeting or sheep fleece around the lift where it cradles the sail bundle and sew it (woolly side out) to the lift through the lay of the line. Space the chafing material so that the lift can be adjusted and so that the sail will still be protected. Alternatively, a nylon strap could be used in this section of the lift. Sew "D" rings in the strap ends to join the eye splices in the rope parts of the lifts.

TABLE 6

ROPE DIAMETER FOR LIFTS

Sail Area (in square feet)	Lift Size (diameter in inches)
100 - 200	¼
200 - 500	⅜
500 - 800	½

The lifts can raise the boom only when a large sail is partly or all the way up. There is not enough purchase to lift the furled sail, but I find that this works fine as long as you remember to haul up the lift before reefing or furling. See Drawings 19 & 20, following pages.

THEORY AND PRACTICE

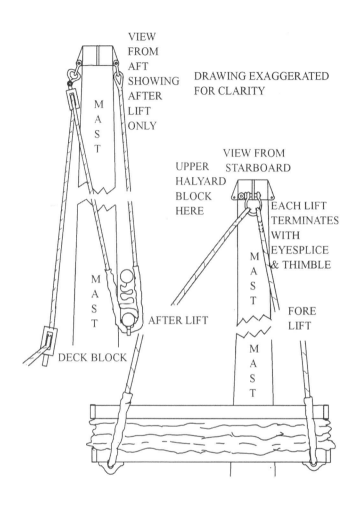

DRAWING 19

LIFTS

THE CHINESE SAILING RIG

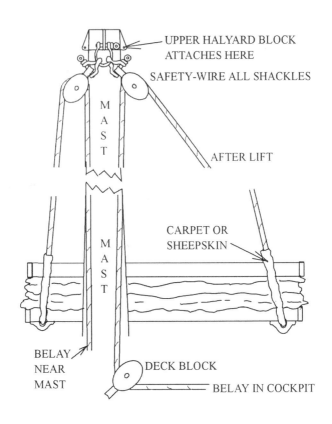

DRAWING 20

LIFTS

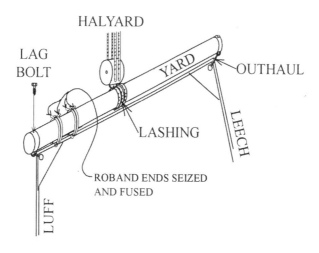

DRAWING 21

YARD RIGGING

PARRELS

The batten parrels, among other things, help to distribute stress along the mast in the same manner as the eyes along a fishing rod. The fish line passing through the eyes bends the rod evenly rather than placing all the force of the line at the end eye. The same is true for the Chinese lug sail; that is, rather than having the whole weight of the sail concentrated at the mast's tip, the mast bends evenly along its length on any point of sailing where the wind pressure on the sail tries to push it away from the mast.

The batten parrels are best made of large diameter polypropylene line because it slides well. They are tied through grommets at the luff and tied to a seizing just aft of the mast at each batten. Plastic garden hose or pipe sections or even wood or plastic beads can be used over smaller parrel line. During the trial sailing period, the batten parrels must be adjustable. Using large diameter (about ¾-inch for a large sail) parrels reduces chafe of both parrels and mast. In addition, the outer rope fibers of large diameter line protect the inner fibers from destructive sunlight. See Drawing 22 facing page.

SHEETS

A question I am frequently asked is, "How do the sheets work?"

There are several ways of rigging Chinese lug sheets. The Chinese traditionally used a "euphroe block." I have sailed with them and found it dangerous to have a chunk of wood flying around in the air attached to the sheets. Also, they make reefing difficult.

The sheeting arrangement I like best was used on sailing canoe rigs during the 1870s and later used by Colonel H. G. Hassler in his Chinese lug designs.

The number of moving blocks on the sail (in other words, the purchase power) depends on the total number of battens, which as I have said, relates to the area of the sail. See Drawing 23, page 74.

THEORY AND PRACTICE

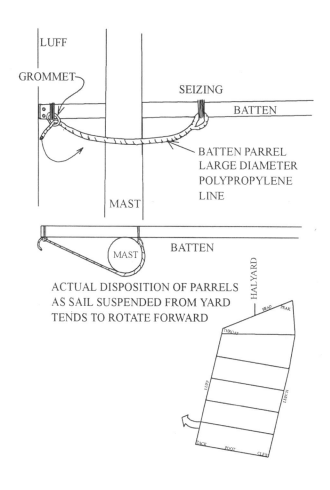

DRAWING 22

BATTEN PARREL ATTACHMENT

65

THE CHINESE SAILING RIG

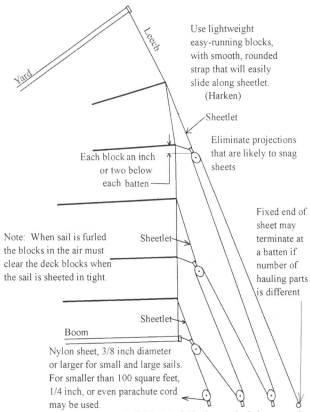

DRAWING 23

SHEETS

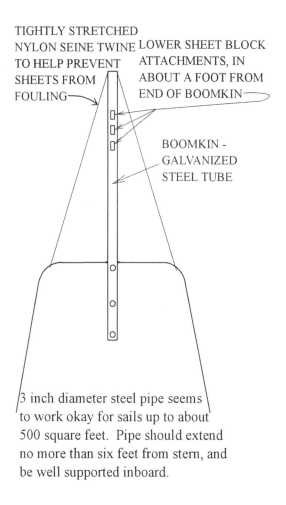

DRAWING 24

A BOOMKIN

THE CHINESE SAILING RIG

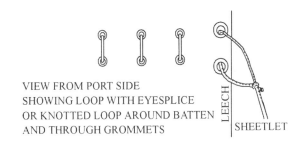

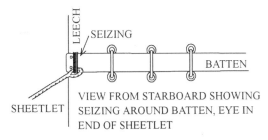

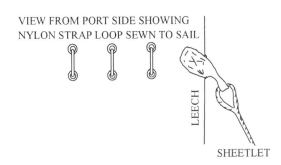

DRAWING 25

THREE METHODS FOR ATTACHING SHEETLETS

TO SAILS

REEFING DOWNHAULS

The batten downhauls that hold the battens down when the sail is reefed are led back to the cockpit. It is not necessary to rig them to all the battens. Only rig as many as you want for the number of possible reefs you want in the sail. The batten downhauls can simply be individual pieces of nylon or Dacron line seized onto the battens at the mast and led aft (see Drawing 26, following page), or they can be rigged as the sheets are, with sheetlets between the battens (seized to the batten at the mast) with the downhaul eye spliced around the sheetlet (with a thimble) and leading aft. See Drawing 27, page 79. This latter method will only work if there is enough distance between the deck block and the bight of the reefing sheetlet when reefed.

The downhaul helps keep the sail from bellying out when reefed and allows luff tension to be set up for sailing to windward.

TABLE 7

DOWNHAUL ROPE DIAMETER

Sail Area (in square feet)	Downhaul Size (in inches)
100 - 200	3/8
200 - 800	1/2

THE CHINESE SAILING RIG

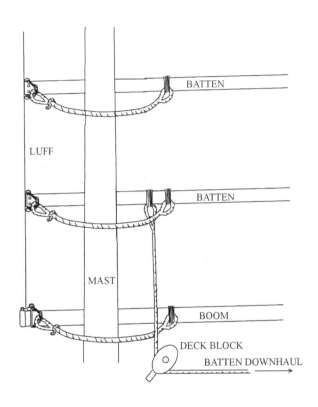

DRAWING 26

SINGLE BATTEN DOWNHAUL

THEORY AND PRACTICE

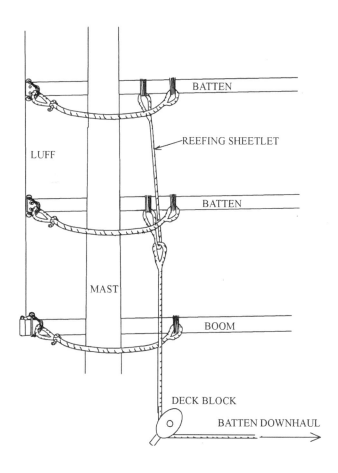

DRAWING 27

TWO-BATTEN DOWNHAUL

Of course, as with any sail, a boom downhaul is needed or the sail will rise up the mast and form a cloth bag. The boom downhaul also helps keep the sail, suspended as it is from one point, from rotating forward. The boom downhaul is adjusted so that it is taut when the sail is fully hoisted.

UNDER WAY

The Chinese Sailing Rig

UNDER WAY

The Chinese lug rig is not much different from a conventional jib headed rig to sail. There is one sheet per sail, one halyard, and one boom lift that needs to be adjusted per sail.

JIBING

There is less danger from an accidental jibe since there is no standing rigging, and the area of sail ahead of the mast opposes the jibe. In fact, in a moderate breeze jibing is accomplished by simply throwing the helm over and letting the jibe happen, taking care to keep out of the way. Although, when the wind is strong it is best to haul and veer the sheet to ease the sail across the vessel, otherwise a "goosewing" jibe may result.

SEEING THE WIND

The main difficulty beginners seem to have in sailing with the Chinese lug rig is in assessing when the sail is drawing or luffing. By watching the sail cloth closely, one can detect areas of tautness and looseness in the cloth between the battens. By observing these areas and sheeting and steering to keep them all just taut, one can obtain optimal performance. Several bright pieces of knitting yarn threaded through the sail, a few inches aft of the luff, provide an excellent indicator of laminar air flow. When the yarns on both sides of the sail are streaming back smoothly, the sail is drawing.

DON'T FORGET

What is most difficult to learn at first is the use of the after lift on a Chinese lug sail. A large sail is heavy and there is not enough purchase in the lift or strength in the crew to raise the sail off the gallows before hoisting. The crew must hoist the sail partially before the lift will get the sail clear of the gallows. When sailing with the sail fully raised, the lift should be slacked off to avoid chafe. The synthetic lift line may continue to "shrink" and become tight again, so check it later.

Now—most important—be sure to set the lift up again before lowering or reefing. Neglect of this may cause the after end of your sail to drop into the water or onto the deck, perhaps ripping the sail. Here I suggest that the gallows be crowned with well-rounded ends. The sail will tend to ride over such a gallows and not be caught as it sweeps across the vessel while being raised or lowered.

EASY REEFING

Reefing is simple. Merely lower as much sail as you wish into the lifts. Adjust the sheet and continue sailing. If you are beating, haul taut the uppermost lowered batten downhaul, then hoist the sail a bit with the halyard, until a taut luff is obtained.

A GOOD IDEA?

Because the mast is tapered, the yard parrel must be loose enough to not jam as the sail is lowered. If you desire you can fit a yard hauling parrel to the yard. This is a length of nylon line attached to the yard near the mast, then passing around the mast it goes through a swivel block or bulls-eye on the yard and down along the mast to a foot block. From the foot block, the yard hauling parrel comes back to the cockpit, where it is belayed. The yard hauling parrel, when set up tightly, keeps the yard from swinging away from the mast and banging back into it. Brian Platt, who wrote about ocean cruising in his junk *High Tea* in the 1960s, concluded that a collar of old fire hose coated with paraffin wax worked best for him. A piece of thick carpeting on the yard and a regular parrel, such as the battens all have, seems to work fairly well, and is simple. With Platt's fire hose, or padding, it is not necessary to remember to slack the yard hauling parrel before lowering sail. See Drawing 28, next page.

The Chinese Sailing Rig

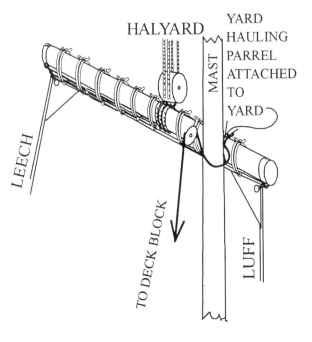

DRAWING 28

YARD HAULING PARREL

SPAGHETTI

The rat's nest of lines in the cockpit area can be organized somewhat by having a box for each line. Don't try to coil the lines into the boxes; just allow them to fall in randomly. The line will pay out better if it is simply allowed to arrange itself. Braided nylon line is pleasant to handle and doesn't kink, but its outer sheath abrades easily. Three-strand line is more durable. Dacron, though more expensive, stretches less than nylon and works best for halyards.

ADVANTAGES

By backing the foresail, hauling the main to weather, and so on, you will find that you can maneuver your Chinese lug rigged vessel into places where boats normally don't go under sail. The ability to lower all sail instantly adds to your confidence in tight places. By alternately backing the fore and the main wing on wing with the wind at the bow, the vessel may be sailed astern by a skipper who handles the helm with care.

The ease with which you can control your Chinese lug sail will soon have you sailing into places where the spectators are dropping their paint brushes and biting their cigars off.

THE DECKED SAILING CANOE RIG

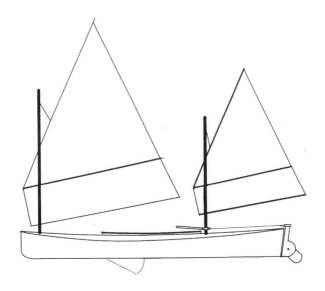

DRAWING 29

VESPER

A TRADITIONAL DECKED SAILING CANOE RIG

THE DECKED SAILING CANOE RIG

HISTORY

The sailing canoe movement of the mid to late 19th century began when John Rob Roy MacGregor had a decked canoe, his adaptation of the skin-covered "Esquimaux kayak," constructed of wood. Though it was primarily a double paddle propelled vessel, he used a small lug sail in fair winds. MacGregor traveled through Europe in his canoe, writing *A Thousand Miles in the "Rob Roy" Canoe on the Lakes and Rivers of Europe.*

Boatbuilders in Europe, Canada, and the United States began to produce these kayak-styled canoes for a growing number of enthusiasts. The paddle-propelled canoe with auxiliary sail evolved into the sailing canoe. J. Henry Rushton of Canton, New York, a prolific boatwright with his own factory that manufactured a wide variety of small vessels, produced *Vesper*, designed by Robert H. Gibson, a member of the Albany, New York, Mohican Canoe Club. *Vesper,* with Gibson at the helm, won the American Canoe Association's International Challenge Cup in 1886. Gibson's canoe was among the last of the camping sailing canoes. Most later designs were devoted to racing.

The sailing canoe, being small and inexpensive, became a great laboratory tool for experimenters. The cam cleat and remote reefing are examples of inventions that came out of the sailing canoe movement. The canoe rig featured easy control from a central position, just what I was after in my larger boats. W. P. Stephens, writing on sailing canoes, described rigs where "... each individual sheet which united with the others in a single main sheet...[permitted the] hauling down two or three reefs in a moment."

MY CANOE

Years ago I was inspired by the ingenuity manifested in the canoe rig. I first gazed upon the sublimely beautiful *Vesper* at the Thousand Islands Shipyard Museum in Clayton, New York. A couple of years ago I built an epoxy/plywood lapstrake version for myself. And so I got to play with the sails.

The canoe rig uses battens, but it differs greatly from the Chinese sail. It is sewn of fabric panels with "broadseams," and with its shape further determined by cutting the perimeter to certain formulated curves. In this I had the help of a publication written by James Lowell Grant, *Make Your Own Mainsails... and Win!* This little book lifts the veil of secrecy from the trade secrets of the professional sailmaker, and enables the amateur to make acceptable Western-style sails. Moreover, Grant's firm, Sailrite, supplies everything needed to construct sails, from the sewing machine to the metal spur grommet.

WHERE TO GET SAILCLOTH, ETC.

Obtaining Dacron sailcloth in small quantities at reasonable prices has gone from difficult to almost impossible. Your local sailmaker may be willing to sell you sailcloth, but to obtain exactly the type of cloth you want, you may have to purchase a bulk lot, leaving you with bolts of leftover cloth. Many amateur sailmakers have been experimenting with plastic tarp sails. I prefer using the real thing, Dacron. Fortunately Sailrite sells sailcloth in any weight I'd ever want, and in a variety of colors.

METHOD

When I construct my canoe sails I first draw them on white-painted plywood. I roll the cloth across the drawing and hot-knife it to size, leaving a couple of inches beyond the perimeter for hemming and for mistakes. The lines on the white-painted plywood are clearly visible through white Dacron sailcloth. Scraps are used for corner patches.

I then roll the bundle of cloth so that it passes under the arm of the sewing machine, and sew the panels together. Next I sew on the corner patches. Batten pockets may also be sewn to the sail at this point in construction. I finish the perimeter with pre-creased Dacron tabling, also sold by Sailrite. Stamped-in brass spur-grommets along the head and foot complete the sail, along with larger, sewn-in corner grommets. I use plastic rings, and pre-bore the holes for the grommet and the stitching with my drill press. I sandwich, and temporarily screw, the cloth between two sheets of thin plywood upon which I've glued a paper hole pattern I've made with a computer drawing program. This makes it much easier to sew the ring into the hole.

FITTINGS

It's not for nothing that decked sailing canoe fittings are referred to as "jewelry." Tiny blocks and cleats seem as though they would be just as appropriately dangling from charm bracelets. Elegant examples are obtainable from specialty sailing canoe hardware suppliers. I find mine at local hardware stores and at my chandlery. For sheet blocks I use nylon rings from West Marine, as well as inexpensive window covering pulleys.

MASTS

I shape my masts from western red cedar, glued up from several planks, arranging them so that the grain makes for compatible planing later. Varnish and paint both suffer from abrasion, so I have had success using Star Brite Tropical Teak Oil. I have made an 18-foot, glued and shaped mast of western red cedar, again using a Star Brite finish. Incidentally, when shaping a spar from squared stock I use the traditional system of eight-siding planing and sanding.

BATTENS

I duplicate the original sail battens in Douglas fir, finishing them with seven coats of varnish. Even though I dislike batten pockets on larger Chinese sails, I decided to use them just as they did in the 1880s because I don't expect to have to remove or replace them at sea.

There is still much to be learned about these sailing rigs, and the decked, wooden sailing canoe. W. P. Stephens called the decked sailing canoe "the poor man's yacht." He praised "the enterprise and ingenuity of the canoeing world" for initiating great progress in the rigging of larger yachts.

QUICK RIG

Small sails can be made in a few hours. I have rigged and re-rigged my 18½-foot Chesapeake Bay crabbing skiff with six different homemade sails.

Here, too, some might choose blue or silver tarp or Tyvek. While I have always sewed the sail panels together with my 1950s-era zigzag machine, a sailmaker friend told me that straight stitching is okay for small sails. Some experimenters use adhesive, but I've not tried it. One-eighths-inch Dacron parachute cord serves for running rigging, except for the halyard, where I employ ¼-inch or thicker rope.

For those interested in experimentation I offer a few suggestions in Drawings 30 and 31 (following pages). These final two drawings suggest shortcuts for those building small rigs. The final photograph, of my crabbing skiff *Syncopation*, shows a Chinese sail with full length batten parrels so that its windsurfer mast may be lowered without removing the sail from the mast.

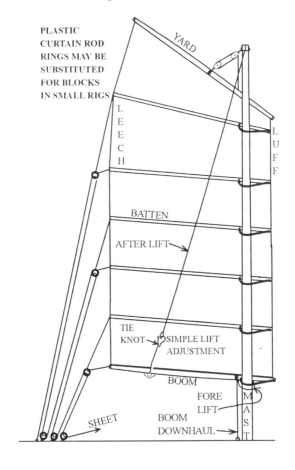

DRAWING 30

QUICK RIG DRAWING

STARBOARD SIDE

THE DECKED SAILING CANOE

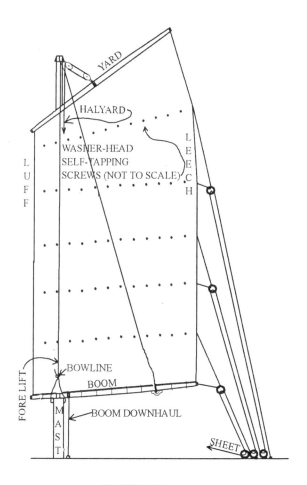

DRAWING 31

QUICK RIG DRAWING

PORT SIDE

Syncopation

PHOTO BY HOWARD VAN LOAN, 2006

READING LIST

Borden, Charles A. (1967). *Sea Quest*. Philadelphia: Macrae Smith Company.
　See the chapter entitled "Ocean Cruising Junks."

Bradshaw, Todd E. (2000). "Canoe Rig—The Essence and the Art." Brooklin, ME: *Wooden Boat*.

Campbell, John. (1986). *Easier Rigs for Safer Cruising*. London: Hollis & Carter.
　Modern junk rigs and beyond.

Chapelle, Howard I. (1936). *American Sailing Craft*. New York: Bonanza Books.
　Examples of unstayed masts in traditional western craft.

Chapelle, Howard I. (1951). *American Small Sailing Craft*. New York: W. W. Norton.
　Traditional working craft under forty feet, with good details.

Chapelle, Howard I. (1964). *Yacht Designing and Planning.* New York: W. W. Norton & Co.

Colvin, Thomas E. (1972). *Coastwise and Offsore Cruising Wrinkles.* New York: Seven Seas Press.

Colvin is a pioneer of Chinese rigged western craft.

Hasler, H. G. & McLeod, J. K. (1988). *Practical Junk Rig.* Camden, ME: International Marine Publishing.

Masters of the art of adapting the Chinese rig to modern craft.

Herreshoff, L. Francis. (1972). *The Compleat Cruiser.* New York: Sheridan House.

Practical advice on cruising vessels with historical data and diagrams.

Hill, Annie. (1993). *Voyaging On A Small Income.* St. Michaels, MD: Tiller.

Pete and Annie Hills' high seas experiences in their 34' Chinese rigged dory.

Hiscock, Eric C. (1965). *Cruising Under Sail.* London: Oxford University Press.

A number of yacht designs of various sizes with all pertinent specifications in these two books plus more for anyone who contemplates sailing.

Hiscock, Eric C. (1965). *Voyaging Under Sail.* London: Oxford University Press.

Macgregor, John. (1867). *A Thousand Miles In The Rob Roy Canoe On Rivers And Lakes Of Europe.* London: Roberts Brothers.

The book that popularized canoeing and kayaking.

Macgregor, John. (1867). *Rob Roy On The Baltic: A Canoe Cruise Through Norway, Sweden, Denmark, Sleswig, Holstein, the North Sea and the Baltic.* London: Sampson Low, Son.

Appendix on design and construction. Great story. May be available online.

March, Edgar J. (1972). *Sailing Drifters.* Camden, ME: International Marine Publishing.

Some big unstayed lug rigs.

Marchaj, C. A. (1964). *Sailing Theory and Practice.* New York: Dodd, Mead & Company.

Aerodynamic analysis well expressed in formulas and graphs.

Manley, Atwood. (1977). *Rushton and His Times in American Canoeing.* The Adirondack Museum/Syracuse University Press.

McMullen, R. T. (1949). Down Channel. London: Rupert Hart Davis.

An amateur designer/sailor of the mid-nineteenth century who experimented with unstayed rigs.

Munroe, Ralph M. & Gilpin. (1966). *The Commodore's Story.* Narberth, PA: Livingston Publishing Company. (1974). Historical Association of Southern Florida.

Monroe, a pioneer Floridian, used unstayed masts and simple work-boat designs. Sailing vessels were his only communication with the outside world for many years.

Needham, Joseph. (1971). *Science and Civilization in China*, Vol. IV:3, Cambridge University Press
Traditional Chinese Rigs.

Parker, Reuel, B. (1994). *The Sharpie Book*. Camden, ME: International Marine.
Making unstayed masts and more.

Ridler, Donald. (1972). *Eric the Red*. London: William Kimber & Co., Limited.
Donald Ridler designed, built, and cruised his Chinese rigged dory, all on a shoestring budget. A great story.

Smith, Hervey Garrett. (1953). *The Arts of the Sailor*. Funk & Wagnalls.
How to do eyesplices, and so forth.

Stephens, William P. (1895). "Canoe And Boat Building—A Complete Manual For Amateurs," *Forest and Stream*. May be available online.

Stephens, William P. (1945). *Traditions and Memories of American Yachting*. New York: Hearst Magazines.
A good historical overview, including a section on batten lug-rigged sailing canoes.

Worcester, G. R. G. (1971). *Junks & Sampans of the Yangtze*. Annapolis, MD: Naval Institute Press.

The bible of junk historians, a masterpiece showing traditional Chinese vessels and the society that produced them.

Worcester, G. R. G. (1966). *Sail and Sweep in China*. London: Her Majesty's Stationery Office.

A condensation of *Junks & Sampans of the Yangtze*.

THE INTERNET

Google: decked wood sailing canoes, junk rigs

www.sailrite.com
    Their online catalogue

email: rblain@sunbirdmarine.com
    Robin Blain is Hon. Sec, The Junk Rig Association

PERIODICALS

Junk Rig Association Newsletter
    Junk Rig Association, Attention Robin Blain, Hon. Sec.
    373 Hunts Pond Road
    Titchfield Common
    Fareham
    Hants
    P014 4PB
    ENGLAND
    All the latest, a treasure-trove of information.

Messing About in Boats
29 Burley Street
Wenham, MA 01924

An all-round good magazine, comes out every two weeks.

GLOSSARY

Airfoil: A device to "fool the air" into propelling a boat or lifting an aircraft. A sail and airplane wing are both airfoils.

Aft, After: Toward the stern.

Aspect Ratio: The relationship between the height of a sail and its breadth. A sail with a height of 10' and 10' of breadth has an A. R. of 1:1; 20' height and 10' breadth A. R. = 2:1, etc.

Batten: A light spar of wood, bamboo, tubular metal or plastic, attached to a sail as a stiffener.

Beat: To sail to windward.

Belay: To secure a line.

Block: A pulley.

Boltrope: A rope sewn to the perimeter of a sail.

Boom: The spar along the foot of a sail.

Boomkin: A spar extending from the stern of a vessel.

Bulkhead: A partition or wall below decks.

Bury (of mast): The distance or depth between the partners and the mast step.

Camber: Curve.

Center of Effort (CE): The geometric center of a sail.

Chafe: Injury to a material due to friction.

Clew: The after lower corner of a fore and aft sail.

Cutter: A vessel with a single mast stepped 4/5 to 1/2 the length of the deck aft of the bow, and setting two or more fore and aft sails.

Downhaul: A line rigged to pull a spar or a sail down.

Fairlead: A device for, or the process of leading a line clear of obstructions.

Foot (of sail): The bottom edge of a sail.

Foresail: The principal sail set on the foremast of a schooner.

Furl (Chinese Sail): To lower sail completely.

Genoa: A large overlapping triangular sail, said to have been first used in a race at Genoa, Italy.

Gollywobbler: A large quadrilateral sail rigged between two masts, the throat setting to the foremast, the peak to the aftermast.

Halyard: Literally a "haul-yard," the line that hauls the yard up the mast.

Head (of sail): The uppermost edge of the sail.

Heel: The leaning or tipping of a vessel to one side or the other.

Hoist (noun): The height of a sail along the mast.

Hoist (verb): To lift, generally with tackle.

Jibe: When sailing before the wind, to bring the wind onto the other side of the sail, so that it swings across the vessel.

Jib Headed: Triangular sail (with a more or less pointed head).

Ketch: A two-masted, generally fore-and-aft rigged vessel with the mizzen ahead of the rudder post.

Knee: An "L" shaped brace of wood or metal.

Lead: The distance from CE to CLR

Lee Helm: A condition in which the tiller (helm) must be held to the lee side of the vessel in order to maintain course. Lee helm is undesirable for safety and for hydrodynamic efficiency.

Leech: The aftermost edge of a sail.

Lift: A line that raises the boom.

Luff: The forward edge of the sail.

Lugsail: A quadrilateral sail, suspended from a yard, with more area aft of the mast than forward.

Mainsail: In a two-masted schooner, the aftermost sail. In a ketch or yawl, the principal forward sail.

Mizzen: The smaller aftermost sail of a ketch, yawl, or three-masted schooner.

Outhaul: A line that pulls a sail out along any spar except the mast.

Partners: The reinforcement at the deck that supports the mast that passes through them.

Parrel: A line that passes around the mast in order to hold a spar to the mast.

Peak: The aftermost upper corner of a quadrilateral sail.

Port Tack (lug rig, square'sl): A condition in which the tack of the sail is on the port side of the vessel.

Rake: Inclination from vertical (generally, fore or aft).

Reef: To reduce the amount of sail exposed to the wind.

Reach: To sail with the wind a-beam.

Robands: Individual lashings that attach a sail to a boom or a yard.

Run: To sail before the wind.

Running Rigging: Lines that move, generally through blocks; distinguished from standing rigging.

Set (noun): Overall quality deriving from the combination of factors that produce a good airfoil such as: smoothness, proper twist, camber, etc.

Set (verb): To raise or hoist sail; to make sail.

Schooner (two masted): A fore-and-aft rigged sailing vessel, with two masts of equal height, or with a shorter foremast.

Seize: To lash with small line or wire.

Sheet (noun): The line used to adjust the sail angle to the wind.

Sheet (verb): The processes of adjusting the angle of the sail to the wind ("to sheet").

Sheetlet: A "little sheet" running between two adjacent battens.

Shroud: A guy from the side of the vessel that supports a mast, bowsprit, or boomkin. A side stay.

Sloop: A single-masted sailing vessel with the mast stepped fairly far forward setting two or more fore and aft sails. (See **Cutter.**)

Spar: A boom, yard, batten, bowsprit, boomkin, gaff. A long stick of wood, tubular metal, or plastic.

Spinnaker: A large, or even huge, triangular sail used to propel a vessel on courses from downwind to reaching. Said to derive from "*Sphinx*'s acre," the yacht *Sphinx* being the first to set such a sail.

Standing Rigging: Wire rope or rope that supports the masts and spars.

Starboard Tack (lug sail, square'sl): A condition in which the tack of the sail is on the starboard side of the vessel.

Stay: Standing rigging used to support a mast, bow sprit, or boomkin.

Step (noun): The device that secures the bottom (foot) of a mast.

Step (verb): The process of placing the mast into a vessel.

Tallow: Grease made from animal fat, used for reducing chafe, protecting wood, and for lubrication.

Throat: The forward upper corner of a quadrilateral sail.

Thwartships: Across the vessel, perpendicular to the keel.

Veer: To let out line.

Weather Helm: A condition in which the tiller (helm) must be held to the windward (weather) side of a sailing vessel in order to maintain course. A slight amount (3 to 7 degrees) of weather helm is desirable for hydrodynamic and safety reasons.

Windward: The side toward the wind.

Yard: A spar that crosses the mast and from which a sail is suspended by its head.